ANIMAL
TRACKS

FROM
MACAULAY
TO
STEPHEN

Distribution of this copy of
ANIMAL TRACKS OF WESTERN CANADA, by Joanne Barwise,
has been undertaken by the
Youth-In-Action program of the
Recreation, Parks and Wildlife Foundation of Alberta. This
book is one of several major works of natural history including
Wild Mammals of Western Canada,
Grizzly Bears, Wolves and *The Wonder of Canadian Birds*
which have previously been distributed to
school libraries by the Foundation.

The Foundation believes the subject and quality of
this book will contribute to our children's understanding of the
fragile beauty of all wildlife, and will help to further the
objectives of the Foundation, which are:

- to develop and maintain recreational programs,
  services and facilities
- to develop and maintain parks
- to manage and conserve fish and wildlife

The Foundation acknowledges with gratitude the
co-operation and assistance provided by
the Honorable Jim Dinning, Minister of Education,
and the Honorable Dr. Stephen West,
Minister of Recreation and Parks,
in making this project possible.

For further information regarding the Foundation, its work,
and how you can participate, contact
The Executive Director
Recreation, Parks and Wildlife Foundation
#705, 10045 - 111 Street
Edmonton, Alberta T5K 1K4
Telephone: (403) 482-6467

ANIMAL TRACKS

OF WESTERN CANADA

Joanne E. Barwise
Illustrated by Ozlem Boyacioglu

PINE
CONE

First printed in 1989 10 9 8 7 6
Printed in Canada

The Publisher:
Lone Pine Publishing,
206, 10426-81 Avenue
Edmonton, Alberta,
Canada T6E 1X5

Canadian Cataloguing in Publication Data

Barwise, Joanne, 1953 -
Animal tracks of Western Canada

Includes bibliographical references.
ISBN 1-919433-20-0

1. Mammals - Canada, Western - Identification.
2. Animal tracks - Canada, Western -
Identification. I. Title.
QL768.B37 1989   599'.09712   C89-091632-2

Cover Illustration: Red Fox by Judi Wild
Editorial: Mary Walters Riskin
Design: Yuet C. Chan
Production: Yuet Chan, Michael Hawkins & Phillip Kennedy
Printing: Quality Color Press Inc., Edmonton, Alberta, Canada

Publisher's Acknowledgement

The publisher gratefully acknowledges the assistance of the Federal Department of Communications, Alberta Culture and Multiculturalism, the Canada Council, the Recreation, Parks and Wildlife Foundation, and the Alberta Foundation for the Literary Arts in the publication of this book.

Canadian Parks and Wilderness Society
Henderson Book Series No. 14
The Henderson Book Series honours the kind and generous support of Mrs. Arthur T. Henderson who made the series possible.

*For my father who taught me
about spruce gum
and my mother who never
stopped me from
climbing trees.*

Acknowledgements

The author gratefully acknowledges the assistance of the Recreation, Parks and Wildlife Foundation in the preparation of this manuscript.

The publisher gratefully acknowledges the assistance of the Federal Department of Communications, Alberta Culture and Multiculturalism, the Canada Council, the Recreation, Parks and Wildlife Foundation, and the Alberta Foundation for the Literary Arts in the publication of this book.

Contents

Welcome to the World of Tracking

Animal Tracks of Western Canada is designed to help you identify the most common mammal tracks found in the western provinces. Reading tracks can be like playing a detective game. The more clues you gather, the more educated your guess will be. If you guess too early, or make a judgement on just one print, your chances of being wrong increase.

Tracks can be found just about anywhere, but most often they are seen along the edges of animals' habitats: where trees and a field meet, or along the edges of ice and open water. Try to avoid stepping on tracks when you are following them. Instead, walk beside them. You may want to go back and look at the track again, or to take some measurements using the guide on the back cover. Furthermore, it's good etiquette to leave the track pattern intact for others to see later.

Track Patterns

All animals, including people, move in a variety of ways. We can run, skip and jump but normally we walk. In the same way, a deer can walk, run, bound or leap, but walking is the normal gait.

Each mammal has its own characteristic gait or movement on foot, and the most common gaits can be classified into four groups: diagonal walkers, bounders, hoppers and pacers. Familiarize yourself with these four basic gaits, and the track patterns they make, by referring to pages 14 to 21.

Identification Pages

Each identification page contains detailed drawings of both front and back feet, or the larger of the feet if they are similar. You'll also find the characteristic track pattern or gait of the animal. When you arrive at the appropriate identification page, compare the measurements you took with the ones provided, and compare the pictures to the actual track. The illustrations are not drawn to scale: the sizes shown are intended to provide a clear comparison for your observations.

An individual animal's print may not always look exactly like the illustration. The drawings are intended as a guide. Make notes with a pencil: they may help you the next time you go tracking.

Happy tracking!

Joanne Barwise

How to Use This Book

1. Look carefully at the track pattern on the ground or in the snow, and select a clear sample of it on which to focus.

   If you've found only one or two clear prints, go to page 24.

   If you've found several clear prints, go to page 25. Start with No. 1 and make a choice, either A or B, that best describes the track pattern you see. (Some descriptions offer you a third choice: C.) Follow the directions closely.

2. Continue choosing the option that best describes the track you are observing until you arrive at the answer, or identification page. If this print does not seem right to you, try again. An example: the odds of seeing a wolf in the city are slim; large, wolf-like tracks found in the city were likely made by a dog.

3. The illustrations of the prints and track patterns found in this book are typical of the ones you are likely to see. All the toes and claws are shown to help those who find clear prints. However, in dry, fluffy snow or on drier ground, the footprint may be just a circle or an oval shape, and not all the toes may show. Taking measurements and identifying the track pattern will usually lead to greater success in the identification of the animal than will the study of a single print.

Abbreviations Used in the Book

LF - left front foot
RF - right front foot
LH - left hind foot
RH - right hind foot

Tracking Terms

Print: The mark or impression made by one foot in the snow or earth. The width is measured from one side of the print to the other. Length is measured from the front to the back of the print; it does not include claw marks.

Track Pattern: A series of four or more single prints or groups of prints made by an animal.

Straddle: The width of the track pattern. It's measured from the outer edges of the prints across the width of the track pattern. It is also called Trail Width.

Stride: The distance between the prints of a walking animal. It's measured from the centre of one print to the centre of the next.

Leap: The distance between pairs of prints (made by bounders) and clusters of prints (made by hoppers). It is measured from the end of the most forward print to the end of the nearest print.

The Four Track Patterns

LF/LH   RF/RH   LF/LH   RF/RH   LF/LH   RF/RH

Diagonal Walkers

This group of animals makes a track pattern that looks almost like a straight line of single prints. When they walk, the limbs on the opposite sides of the body move at the same time. As the left front foot moves forward, so does the right rear foot. The hind feet land in the same prints made by the front feet. The prints you see in this pattern are almost always the back feet, although sometimes the hind foot lands beside the front foot, or a little behind it.

or

When these animals move faster, different track patterns appear:

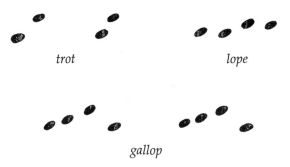

trot   lope

gallop

This pattern is common to the two-toed animals (like deer, moose, elk, antelope, mountain goats and bighorn sheep), as well as to felines (domestic and wild cats) and canines (domestic and wild dogs).

| | LF/LH | | LF/LH |
| | RF/RH | | RF/RH |

Bounders

This track pattern looks like evenly-spaced pairs of prints. These animals have short legs and long bodies. When they bound, they reach out with both front feet and bring both back feet to land just behind where the front feet landed. In dry conditions, the overlapping front and rear prints may look like a single pair of prints.

When these animals move slower or faster, their gait changes to one of the following pattern variations:

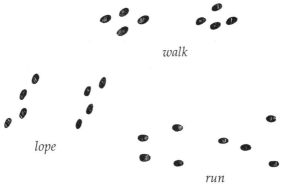

walk

lope

run

This pattern is common to most members of the weasel family. An exception is the slow-moving skunk, which tends to leave a meandering trail.

Hoppers

This track pattern shows clusters of prints - two small and two large. When these animals move, they push off with their hind feet and land with their front feet. The hind feet swing all the way through to land in front of the front feet. The gait is common among animals with large back feet. The pattern usually remains the same no matter how fast or how slow the animal goes, although you may see some variation, such as the one shown here:

This pattern is common to all rabbits, hares and most rodents, with the exception of porcupines, beavers and muskrats which are Pacers.

If you noticed the snowshoe hare in the upper right hand corner of the odd-numbered pages, then you have the observation skills to make a good tracking detective. Flip the corners and watch the hare hop. You may be fortunate to see the real thing when you are out exploring.

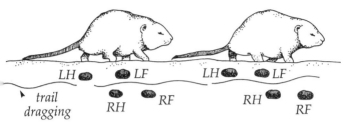

trail dragging

LH   LF   LH   LF

RH   RF   RH   RF

Pacers

Heavyset animals make this pattern. They waddle
from side to side creating a track pattern where
each foot makes its own print. Their hind feet are
always larger than the front feet. They move both
limbs on one side of the body and then move both
limbs on the other side. Sometimes the smaller front
foot appears beside the hind print like this:

20

or the hind foot lands on top of, or overlaps, the
front foot like this:

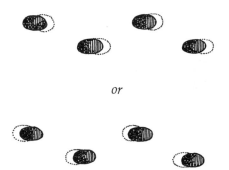

or

This pattern is made by all the wide-bodied
animals, such as beaver, bear, raccoon, muskrat,
skunk and badger.

Tracking Tips

One or two clear prints can sometimes lead to a successful identification, but not always. Try to identify the *track pattern* so you can start at the beginning of the choices in this book on page 25.

Jumping animals that live in trees (squirrels, deer mice) usually leave a print with the smaller front feet together.

direction of travel

Jumping animals that live on the ground (hares, rabbits, shrews) usually put their smaller front feet down, one in front of the other.

direction of travel

Felines (members of the cat family) never leave claw marks and have a round print like this:

Canines (members of the dog family) leave claw marks and have an oval print like this:

Most hoofed animals (deer, moose, sheep, goats, elk) have two large toes or hoof segments on each foot. Their toes spread when they run. The dew claws may leave a mark in snow, like two small circles behind the larger toes. The antelope does not have dew claws.

dew claws

Rabbits and hares leave a branch cut neatly at approximately a 45° angle when they bite off buds.

Deer and moose tear buds off with their teeth, leaving the end of a branch ragged and ripped.

Look for other clues and signs that the animal may have left behind, such as tail marks, claw marks near the toes, bits of hair or fur, tunnels in the snow, or gnawed branches.

Identification Key

One Clear Print

If you cannot identify the track pattern but have a
clear impression of one footprint that has . . .

 two large toes (or hoof):
Go to No. 4.

 four toes around a central pad, an oval
print that is longer than it is wide and
claw marks usually visible:
Go to No. 5.

 four toes around a central pad, a
circular print that is as long as it is
wide. No claw marks:
Go to No. 5.

 A long heel mark or a large back foot
that generally shows no toe pads:
Go to No. 15.

 five oval toes arranged across the top
of the front pad (the fifth toe may not
show), claw marks usually showing:
Go to No. 18.

 long thin toes that look like little hands:
Go to No. 24 and No. 25.

Starting with a Good Track Pattern

1. **A.** If the track pattern looks almost like a
    straight line of prints

Go to No. 4

OR

    B. If the track pattern consist of evenly-
    spaced pairs OR clusters of prints OR a
    large print is paired with a small print

       like this

       this

       or this

Go to No. 2

2. **A.** If the track pattern has evenly-spaced pairs of prints that are the same size

Go to No. 17

OR

B. If the track pattern shows large back feet and small front feet

like this

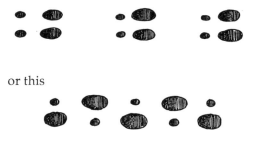

or this

Go to No. 3

3. **A.** If the track pattern has two small prints and two large prints per cluster like this

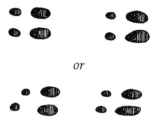

or

Go to No 15

OR

B. If the track pattern has one small print and one large print per pair of prints

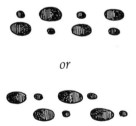

or

Go to No. 21

4. **A.** If each print has two large toes (or hoof)

Go to No. 10

OR

B. If each print has four toes arranged around a pad, with a round or oval shape

 *or*

Go to No. 5

5. A. If the print is an oval shape
        and claw marks show in
        front of the toe pads

Go to No. 8

OR

B. If the print is a round shape
    and no claw marks show in
    front of the toe pads

Go to No. 6

6. A. If the print is approximately
        2 to 3 cm long and 2 to 3 cm
        wide

Go to Page 45

OR

B. If the print is larger than 3
    cm long and 3 cm wide

Go to No. 7

7. **A.** If each print is approximately 5 cm long and 5 cm wide

Go to Page 47

OR

B. If each print is approximately 8 cm long and 8 cm wide

Go to Page 49

OR

C. If each print is approximately 10 cm long and 10 cm wide

Go to Page 51

8. **A.** If the track pattern is almost a straight line with a straddle measuring 7.5 to 10 cm

 7.5 to 10 cm

Go to Page 53

OR

B. If the track pattern is one of a diagonal walker with a straddle measuring from 10 to 15 cm

10 to 15 cm

Go to No. 9

9. **A.** If the print is approximately 6.5 cm long and 5.5. cm wide

Go to Page 55

OR

B. If the print is approximately 11 cm long and 10 cm wide

Go to Page 57

31

10. **A.** If each print is shaped like a split heart

Go to No. 11

OR

B. If each print is shaped like a split pear

Go to No. 14

11. **A.** If the print is heart-shaped and measures 9 cm or less

Go to No. 12

OR

B. If the print is heart-shaped and measures 12 cm or more

Go to No. 13

12. **A.** If the print is 6.5 to 7 cm long

Go to Pages 59, 61

OR

B. If the print is 8 to 9 cm long

Go to Page 63

13. **A.** If the print is approximately 12 cm long

Go to Page 65

OR

B. If the print is approximately 16 cm long

Go to Page 67

14. A. If the print has two toes (or hoof) and is approximately 6.5 cm long

Go to Page 69

OR

B. If the print has two toes (or hoof) and is approximately 7.5 cm long

Go to Page 71

15. A. If the prints of the large feet are approximately 2 cm long or less

Go to No. 16

OR

B. If the prints of the large feet are approximately 5 cm long

Go to Page 73

OR

15. **C.** If the prints of the large feet are approximately 8 cm long or more

Go to Pages 75, 77, 79

16. **A.** If the leap measurement averages 5 cm or less between the clusters of prints

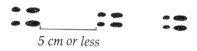

5 cm or less

Go to Page 81

OR

B. If the leap measurement averages 6 cm or more between the clusters of prints

6 cm or more

Go to Page 83

17. **A.** If the average leap measurement between these very tiny double prints is 10 cm or less

10 cm or less

Go to Page 85

OR

B. If the average leap measurement between the pairs of prints is 15 cm or more

15 cm or more

Go to No. 18

18. **A.** If the trail width or straddle measures 6 to 8 cm (five toes may show; in winter, tracks often disappear into snow tunnels)

6 to 8 cm

Go to No. 19

OR

B. If the trail width or straddle measures from 9 to 26 cm (five toes may show)

9 to 26 cm

Go to No. 20

19. A. If the prints are round (toe pads may not show) measuring approximately 1.5 to 2 cm long and 1.5 to 2 cm wide

Go to Pages 89, 90, 91

OR

B. If the prints are round and measure approximately 3.5 cm long and 3.5 cm wide

Go to Page 87

20. A. If the prints measure 3.5 cm long and 4 cm wide

Go to Page 93

OR

B. If the prints measure 7 cm long and 6.5 cm wide

Go to Page 95

OR

C. If the prints measure 9 cm long and 10.5 cm wide (webbing may show between the toes)

Go to Page 97

21. A. If the track pattern is in the bottom of a trough or shallow channel made by the animal's body in the snow, and the prints are toed-in

Go to No. 23

OR

B. If the prints are not in a trough

or

Go to No. 22

22. **A.** If the prints are very large and the bigger foot is in front of or on top of the smaller foot

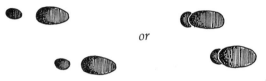

or

Go to Page 101, 103

OR

B. If the track pattern shows the small print beside the large print

Go to No. 24

OR

C. If the track pattern shows the small print in front of the large print

Go to No. 25

23. A. If the track pattern shows drag or scuff marks after each print, especially in snow, and the straddle measures 20 cm or more

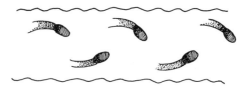

Go to Page 99

OR

B. If the track pattern shows no drag marks and claw marks show on the front feet

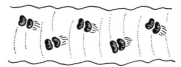

Go to Page 105

24. **A.** If the larger print shows webbed feet and claw marks

Go to Page 107

OR

B. If the small print is hand-like with long nails and the large back foot is not webbed

Go to Page 109

25. A. If the tracks are slightly toed-in with claw marks showing on the smaller front foot

Go to Page 111

OR

B. If the large foot is placed behind the small foot and the tail leaves a drag mark

Go to Page 113

The Mammals

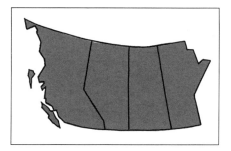

44

House Cat
Felis catus

House cats come in a variety of
breeds, colours and sizes but their
tracks are all about the same size, and
are too small to be confused with
those of other felines. They walk
neatly and leave a fairly straight trail
of prints. Domestic cats are excellent
hunters and will catch birds and small
mammals.

Straddle: 7.5 - 8 cm
   Stride: 12 - 20 cm
     Print: 2 - 3 cm long
            2 - 3 cm wide

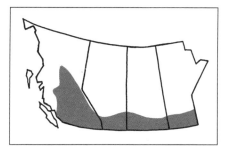

Bobcat
Lynx rufus

The bobcat or wildcat is chiefly nocturnal so it is rarely seen. It is a secretive animal and an excellent stalker and hider. Its tracks are distinctive: round, with four toes and no claws showing, and twice the size of a domestic cat's tracks. The bobcat eats mainly rabbits and rodents.

Straddle: 11 - 12 cm
Stride: 25 - 35 cm
Print: 4.5 cm long
4.5 cm wide

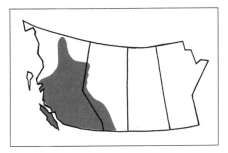

Cougar
Felis concolor

Cougars are found mainly in the mountains and foothills. They are more active at night than during the day, and they rely more on hearing and sight than on smell for locating their prey. They hunt primarily deer and elk, although they will also eat smaller mammals and birds. In deep snow, cougars may leave marks where their feet have dragged on the surface, and occasionally may leave tail marks. The cougar is also called puma, panther and mountain lion.

Straddle: 20 cm
Stride: 40 - 50 cm
Print: 8 cm long, 8 cm wide

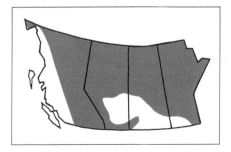

Canada Lynx
Lynx canadensis

The feet of the lynx are so heavily
furred in winter that the track looks
round and the toe pads do not show.
The population of lynx in a particular
year is closely related to the number of
snowshoe hares, which are their main
food. Lynx inhabit densely wooded
areas.

Straddle: 18 cm
   Stride: 30 - 36 cm
    Print: 10 cm long
           10 cm wide

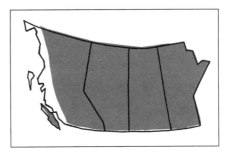

Red Fox
Vulpes vulpes

The trail a fox makes is a single, neat, straight line which usually follows aspects of the landscape such as fences, or the edges of fields and woods. The fox's pads are smaller than those of a dog, and the foot is longer and narrower. In a clear print, the heel pad will show an inverted V-shaped, calloused ridge across the pad, which is not found in other canines.

Straddle: 8 - 10.5 cm
Stride: 30 - 40 cm
Print: 5.5 cm long
5 cm wide

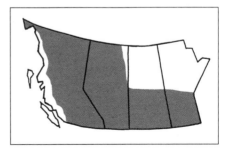

54

Coyote
Canis latrans

Coyote tracks can be easily confused with those of the fox or the domestic dog. Dogs tend to wander, while coyotes will travel with purpose and cunning and will boldly cross through open fields. Coyotes are extremely adaptable animals and they have adjusted to human changes to their habitats. They can run at speeds up to 50 km an hour. They hunt in packs, and prefer the open rangeland.

Straddle: 10.5 - 15 cm
Stride: 30 - 40 cm
Print: 6.5 cm long
5.5 cm wide

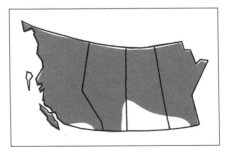

Gray Wolf
Canis lupus

Wolves usually travel in packs of up to twelve individuals. To save energy when they travel in the winter, members of the pack follow in the exact trail made in the snow by the leader. Some domestic dogs will have tracks the same size as the wolf's. A wolf will approach a new object downwind in a circle, while a dog will approach it directly, without caution.

Straddle: 16 - 18 cm
Stride: 40 - 46 cm
Print: 11 cm long
10 cm wide

Inset: *Mule deer bounding*

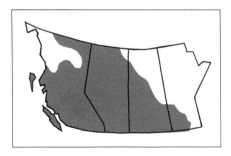

Mule Deer
Odocoileus hemionus

Mule deer have large, mule-like ears and black-tipped tails. When they run, they have a distinct, bounding gait where all four feet come down together — as if the deer were on springs. Their bounding stride can measure between 4.5 and 6 metres. It is difficult to tell the difference between their normal walking tracks and those of the white-tailed deer.

Straddle: 15.5 cm
Stride: 50 - 61 cm
Print: 8.5 cm long
6.5 cm wide

mule deer bounding

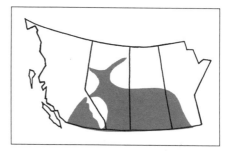

White-tailed Deer
Odocoilius virginianus

In winter, the white-tailed deer's coat
changes from a reddish brown to gray.
Sometimes the deer will drag its feet
between steps, and the track pattern
will appear to be two narrow troughs.
At full speed, the stride increases 3 to
4.5 metres; the print shows spread
toes and dew claws.

Straddle: 15.5 cm
    Stride: 33 - 50 cm
    Print: 7 - 9 cm long
             4.5 - 6.5 cm wide

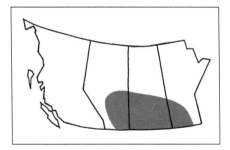

Pronghorn Antelope
Antilocapra americana

Pronghorn tracks resemble those of the deer, but pronghorn antelopes have no dew claws. They are the fleetest of North American mammals, reaching speeds of 80 km an hour. The pronghorn has a habit of scraping the ground with a hoof before depositing urine or droppings on the bare ground. Both sexes have antlers that are shed each fall.

Straddle: 9 - 15 cm
Stride: 30 - 50 cm
Print: 6.5 cm long
5.5 cm wide

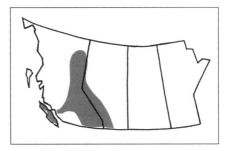

Elk
Cervus elaphus

Elk tracks are larger and rounder than those of deer and are rounder and a little smaller than those of moose. Elk are fond of the bark on aspen trees and during the winter will gnaw the bark with their lower teeth. After years of this, the bark becomes rough and blackened. Look for this sign in the habitat range of the elk. Elk are also called wapiti.

Straddle: 20 cm
Stride: 65 - 70 cm
Print: 10 cm long
7.5 cm wide

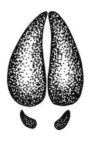

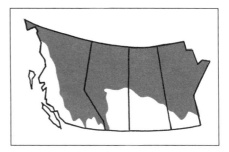

Moose
Alces alces

Moose tracks look similar to those of deer but are twice as large. The large dew claws help to spread the moose's weight over a greater surface area and prints of these claws can be seen where hoof prints are 3 cm deep or more. Because moose drag their feet in deep snow, the track pattern there will look like two narrow troughs. Moose tracks are common around the edges of water bodies where willow, one of the moose's favorite foods, often grows.

Straddle: 23 - 26 cm
Stride: 60 - 85 cm
Print: 16 cm long, 14 cm wide

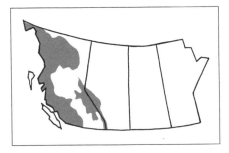

Mountain Goat
Oreamnos americanus

Mountain goats are excellent mountaineers. Their tracks resemble those of the bighorn sheep, because their toes spread to make a square-shaped print. Deep snow will force the goats down from the mountain peaks to search for food. They bed down on rocky ledges and will take refuge in caves. In winter they have thick fur coats; of the 15 to 20 cm of outer guard hairs, 10 cm is a layer of wool.

Straddle: 16 - 18 cm
Stride: 38 cm
Print: 6.5 cm long
       3.5 cm wide

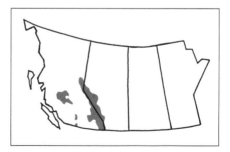

Bighorn Sheep
Ovis canadensis

Bighorn sheep are creatures of regular habits. They often bed down in the same place each night, and will use caves for protection against bad weather. The hooves have straighter edges than those of the deer, and they are not heart-shaped. There is a slight hollow on the inside of each hoof. Bighorn sheep will move from mountain meadows and the foothills to lower elevations during the winter.

Straddle: 17 - 20 cm
Stride: 45 - 58 cm
Print: 7.5 cm long
5 cm wide

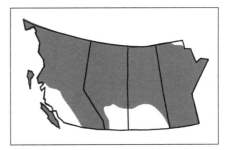

Red Squirrel
Tamiasciurus hudsonicus

Squirrel tracks are commonly seen between one tree and another. Squirrels usually place their front feet next to each other, rather than putting one slightly in front of the other. Their tracks will be found near small holes where they have dug up cones buried in the ground. Cone scales scattered on a stump or log show where the squirrel has stopped to eat the seeds. In deep snow, the tracks look like two diamond shapes.

Straddle: 8 - 12 cm
Stride: 13 - 25 cm
Print: Front - 3 cm long
3 cm wide
Hind - 4.5 cm long
2.5 cm wide

*track impressions
in deep snow*

73

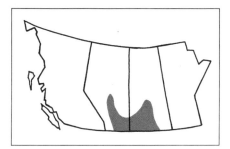

Nuttall's Cottontail
Sylvilagus nuttallii

Cottontails do not change colour to match the season as hares do. During severe winter conditions, cottontails allow themselves to be buried by falling snow and go into a drowsy sleep. When the weather clears, they dig themselves out again. When they run, their leap increases to 0.4 to 1 metre.

Straddle: 10 - 15 cm
  Stride: 18 - 30 cm
    Print: Front - 2.5 cm long,
               2 cm wide
          Hind - 8 cm long,
               2.5 cm wide

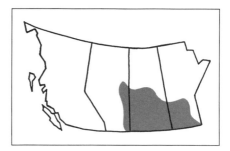

White-tailed Jack Rabbit
Lepus townsendii

The jack rabbit is bigger than the snowshoe hare with larger ears and a longer white tail. During the winter its coat is white and the ear tips are black. If the snow is deep enough, it will dig a shallow hole in the snow where it can protect itself from the weather.

Straddle: 20 cm
Stride: 25 - 40 cm
Print: Front - 5 cm long,
           4 cm wide
      Hind - 15.5 cm long,
           4 cm wide

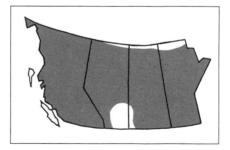

Snowshoe Hare
Lepus americanus

The snowshoe hare is also known as the varying hare because it is brown in the summer but turns white during the winter. The ear tips are black. Hares have distinctive large back feet which adapt them well for travel on snow. When they run, their leap increases to 0.5 to 2 metres.

Straddle: 18 - 20 cm
Stride: 25 - 30 cm
Print: Front - 4.5 cm long
4 cm wide
Hind - 13 cm long
9 cm wide

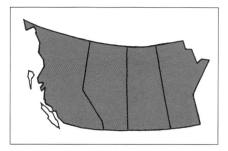

Masked Shrew
Sorex cinereus

Shrews live frenzied lives. It is difficult to see these animals because of their small size, secretive ways, and nocturnal habits. They make runways under the snow, so look for a slight ridging on the snow's surface, or the small entry holes to their tunnels. Shrews also make tunnels under leaf litter and through grasses, like voles do, 2.5 cm diameter or less. Sometimes their tails will leave marks in the snow with their footprints.

Straddle: 2 cm
  Stride: 3 - 5 cm
    Print: Front - .5 cm long,
                .5 cm wide
             Hind - 15 cm long,
                .7 cm wide

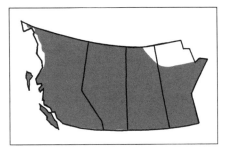

Deer Mouse
Peromyscus maniculatus

Deer mice are great seed-eaters, usually storing more than a litre of seeds and nuts for winter. The food cache is away from the nest, which may be made in a burrow in the ground or built above the ground. Sometimes deer mice renovate old birds' nests in shrubs by making a dome over the nest with shredded materials. Although the tracks of their front feet are paired beside each other, indicating a tree climber, they spend most of their time on the ground. When the snow is soft and fluffy, a tail mark may show.

Straddle: 3.5 - 5 cm
Stride: 5 - 13 cm
Print: Front - .6 cm long
.6 cm wide
Hind - 1.5 cm long
1.0 cm wide

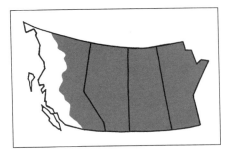

Meadow Vole
Microtus pennsylvanicus

Voles travel through meandering tunnels about 3.5 cm in diameter for protection from the owls, hawks, foxes, coyotes and bobcats that prey on them. They tend to live in marshy areas and grassy fields, and the average vole lives only 2 to 3 months. Tracks are similar to the those of the deer mouse, and occasionally the tail will leave a mark.

Straddle: 3 cm
   Stride: 1.3 - 4 cm
    Print: Front - 1.3 cm long
                    1.3 cm wide
            Hind - 1.5 cm long
                    2.5 cm wide

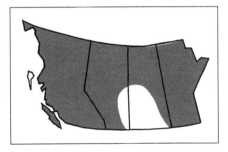

Mink
Mustela vison

Mink have five toes on each foot but only four may be visible; a tail mark may be present in deep snow. Mink make their dens in burrows near water, and usually hunt on the edges between land and water. Their main food is the muskrat. They are less skilled as hunters than their relatives the weasel and the otter. Like the otter, mink will slide on snow, leaving a trough mark.

Straddle: 6 - 8 cm
Stride: 22 - 50 cm
Print: 3.5 cm long
3.5 cm wide

Weasels burrow beneath the snow and will often make right-angled turns. Sometimes the body will leave a mark connecting one pair of prints to another. Weasel tracks are similar to mink tracks, but the straddle and track size are substantially smaller. The fifth toe doesn't always show. The fur of all three weasels changes from reddish brown to white in preparation for winter.

Ermine
Mustela erminea

During the winter, ermine have white feet and can be identified by the black tip on the tail. Formerly known as the short-tailed weasel, ermine is now the more common name for this animal.

Straddle: 8 cm
Stride: 30 - 100 cm
Print: 1.8 cm long
1.8 cm wide

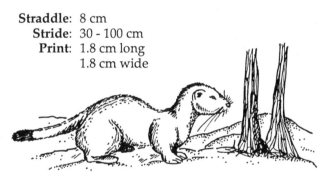

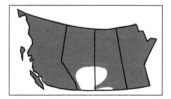

Least Weasel
Mustela nivalis

The least weasel is the smallest of the weasels and is half the size of the long-tailed weasel. It has no black tip on its tail, which is short, and in winter it has white feet. Even though this weasel has a wide distribution over much of the western provinces, it is rather rare and is found in isolated areas.

Straddle: 4 cm
Stride: 20 - 30 cm
Print: 1.2 cm long
1 cm wide

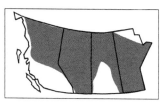

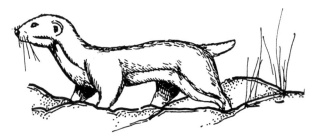

Long-tailed Weasel
Mustela frenata

It is difficult to tell the difference between the long-tailed weasel's and the ermine's prints because their track measurements are similar. When this weasel runs, its black-tipped tail stands up straight. It has brown feet in summer, which change to white in winter.

Straddle: 8 cm
Stride: 15 - 130 cm
Print: 2 cm long
2 cm wide

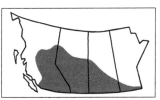

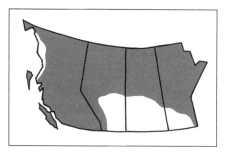

Marten
Martes americana

The marten is a shy animal resembling the weasel. It keeps to remote areas of the wilderness where only trappers may go. Although mink and marten tracks are very similar, habitat usually provides the answer in identification of a track — the marten prefers dry land; the mink prefers water. The marten climbs trees as well as squirrels do. In midwinter its feet are heavily covered with hair, so the toe pads may not be visible in snow. Towards the end of winter, the toes appear in the print.

Straddle: 9 - 12 cm
Stride: 25 - 65 cm
Print: 3.5 cm long
4 cm wide

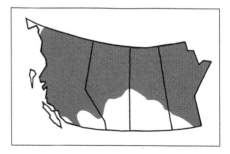

Fisher
Martes pennanti

The prints of the fisher are similar to those of the mink but are twice as large. The toes appear pointed because the claws are close to the toe pads. Prints are usually seen in forested wilderness areas, where fishers den in trees or on the ground. Fishers are excellent climbers, but not as good as the martens. A fisher is also called a pekan.

Straddle: 15 - 20 cm
Stride: 30 - 45 cm
Print: 7 cm long
6.5 cm wide

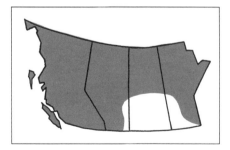

River Otter
Lutra canadensis

Otters like to slide whenever they find it easier than bounding through the snow. The slides begin and end with prints, as otters seem to depend on the force of a leap to propel their bodies. They spend most of their time near the water, but travel overland for food and to reach open water. They are playful, and will slide downhill just for the fun of it.

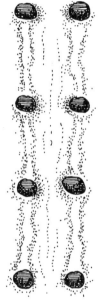

Straddle: 21 - 25 cm
   Stride: 33 - 46 cm
    Print: Front - 6.5 cm long
                   7.5 cm wide
           Hind - 9 cm long
                   10.5 cm wide

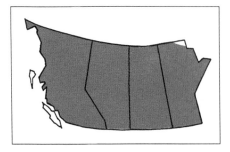

Porcupine
Erethizon dorsatum

The porcupine has a distinct, waddling gait with pigeon-toed prints. As it lumbers along, it leaves a trough in the snow or drag marks on the ground. Its quilled tail leaves a faint criss-cross pattern as it brushes back and forth between the prints. In the track pattern, the back feet usually land ahead of the front feet of the previous step. In shallow snow, the feet leave drag marks. Look for feeding signs of the porcupine in trees, where they like the soft inner bark, buds and twigs.

Straddle: 20 - 24 cm
Stride: 13 - 16 cm
Print: Front - 7 cm long, 4 cm wide
Hind - 11 cm long, 4 cm wide

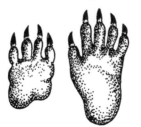

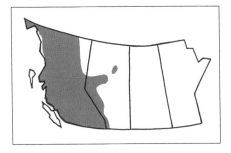

Grizzly Bear
Ursus arctos

The grizzly bear is found in remote
mountainous and foothill terrain. Its
tracks will show claw marks while
those of the black bear may not; the
claws are twice as long as the toe pads.
Grizzlies are inactive in winter, but
their tracks may show around March.
They habitually use the same trails.
For your safety, leave the area if you
see these prints.

Straddle: 35 cm
Stride: 31 - 45 cm
Print: Front - 11.5 cm long
10 cm wide
Hind - 18 cm long
9 cm wide

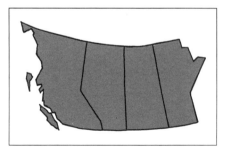

Black Bear
Ursus americanus

The claw marks of the black bear are usually the length of the toe pad. The bear walks in a pigeon-toed manner, where the tracks of the hind feet overlap those of the front feet. The long heel on the back foot makes the print look almost human. Black bears go into a period of dormancy during the winter, emerging from their dens in early spring.

Straddle: 46 cm
Stride: 46 - 50 cm
Print: Front - 14 cm long
              13 cm wide
          Hind - 25 cm long
              14 cm wide

3 cm

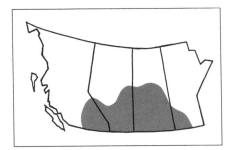

Badger
Taxidea taxus

Badgers are extremely pigeon-toed.
They are well adapted for digging,
with powerful limbs and long claws
which show on their front foot prints.
They do not hibernate in winter, but
they do sleep below the frost line and
will emerge into open fields from their
underground dens to hunt. You may
see numerous holes with fresh prints
nearby them in open grasslands and
on the prairie. The short-legged body
of the badger leaves a trough in the
snow.

Straddle: 10 - 18 cm
Stride: 16 - 30 cm
Print: Front - 6.5 cm long, 5 cm wide
Hind - 5 cm long, 5 cm wide

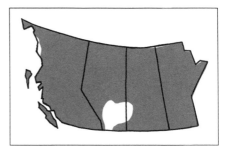

Beaver
Castor canadensis

The track pattern of the beaver shows a front footprint beside a back foot print, which has narrow heel and large webbed toes. The tail drags and often partially or completely covers the footprints. In the snow, beavers leave a trough similar to the porcupine's, but their webbed feet are distinctive. Beaver tracks are rarely seen in winter but are seen occasionally in early spring snowfalls and are more common later in the spring, and during the summer and autumn.

Straddle: 15 - 21 cm
Stride: 10 - 16 cm
Print: Front - 8 cm long, 7 cm wide
Hind - 14 cm long, 12 cm wide

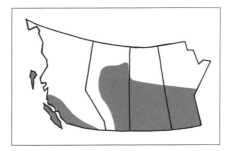

Raccoon
Procyon lotor

The prints of the raccoon are usually paired, with a hind foot beside a front foot. The prints look like those of a small child. Since raccoons are tree climbers, they have long claws which may be seen in the print. Look for the raccoon's prints primarily in spring, summer and autumn. They sleep for long periods during the winter, but do come out looking for food during warm spells.

Straddle: 9 - 13 cm
   Stride: 15 - 40 cm
    Print: Front - 7 cm long
                     7 cm wide
           Hind - 10 cm long
                     6 cm wide

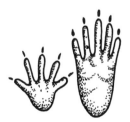

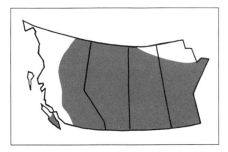

Striped Skunk
Mephitis mephitis

No other animal makes a track pattern like the skunk's. The front feet prints are slightly pigeon-toed and are placed just ahead of the prints of the back feet. The prints may show the long, digging claws. The skunk is a slow-moving animal that leaves a distinct meandering trail. Its tracks are usually found around the den. Inactive in winter, skunks will venture out during warm spells and appear with increasing frequency as spring begins.

Straddle: 7 - 10 cm
Stride: 10 - 20 cm
Print: Front - 3.5 cm long
3 cm wide
Hind - 5 cm long, 3.5 cm wide

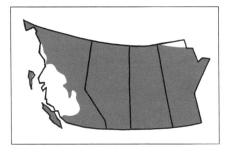

Muskrat
Ondatra zibethicus

Muskrat tracks are found near water and are easy to recognize by the wavy line between the prints made by the long tail. The tracks overlap, with the back print covering a portion of the front print. A muskrat looks very similar to a beaver in the winter, except that you will see its thin tail when it swims. A beaver's tail is not visible when the animal is swimming. A beaver is about four times bigger than a muskrat.

Straddle: 9 cm
Stride: 8 - 15 cm
Print: Front - 4 cm long, 5 cm wide
Hind - 8 cm long
5.5 cm wide

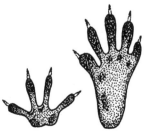

Mammals Listed by Families with Common and Scientific Names

Insect-eaters: Order *Insectivora*
  Family *Soricidae*
    Masked Shrew - *Sorex cinereus*

Hares and Rabbits: *Order Lagomorpha*
  Family *Leporidae*
    Nuttall's Cottontail - *Sylvilagus nuttallii*
    Snowshoe Hare - *Lepus americanus*
    White-tailed Jack Rabbit - *Lepus townsendii*

Gnawing Mammals: Order *Rodentia*
  Family *Sciuridae*
    Red Squirrel - *Tamiasciurus hudsonicus*
  Family *Castoridae*
    Beaver - *Castor canadensis*
  Family *Cricetidae*
    Deer Mouse - *Peromyscus maniculatus*
    Muskrat - *Ondatra zibethicus*
    Meadow Vole - *Microtus pennsylvanicus*
  Family *Erethizontidae*
    Porcupine - *Erithezon dorsatum*

Meat-Eaters: Order *Carnivora*
  Family *Canidae*
    Coyote - *Canis latrans*
    Gray Wolf - *Canis lupus*
    Red Fox - *Vulpes vulpes*
  Family *Ursidae*
    Black Bear - *Ursus americanus*
    Grizzly Bear - *Ursus arctos*

Family *Procyonidae*
  Raccoon - *Procyon lotor*
Family *Mustelidae*
  Marten - *Martes americana*
  Fisher - *Martes pennanti*
  Ermine - *Mustela erminea*
  Long-tailed Weasel - *Mustela frenata*
  Least Weasel - *Mustela nivalis*
  Mink - *Mustela vison*
  Badger - *Taxidea taxus*
  Striped Skunk - *Mephitis mephitis*
  River Otter - *Lutra canadensis*
Family *Felidae*
  House Cat - *Felis catus*
  Cougar - *Felis Concolor*
  Lynx - *Lynx canadensis*
  Bobcat - *Lynx rufus*

Even-toed Hoofed Mammals: Order *Artiodactyla*
  Family *Cervidae*
    Mule Deer - *Odocoileus hemionus*
    White-tailed Deer - *Odocoileus virginianus*
    Moose - *Alces alces*
    Elk - *Cervus elaphus*
  Family *Antilocapridae*
    Pronghorn Antelope - *Antilocapra americana*
  Family *Bovidae*
    Mountain Sheep - *Oreamnos americanus*
    Bighorn Sheep - *Ovis canadensis*

What You Might See and Where

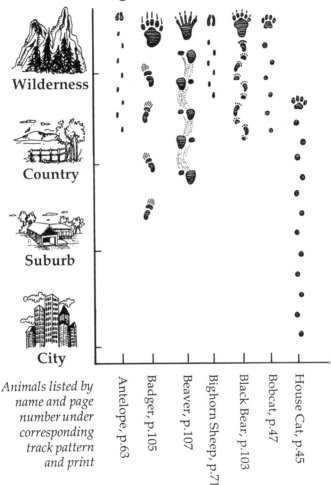

Wilderness

Country

Suburb

City

Animals listed by name and page number under corresponding track pattern and print

Antelope, p.63

Badger, p.105

Beaver, p.107

Bighorn Sheep, p.71

Black Bear, p.103

Bobcat, p.47

House Cat, p.45

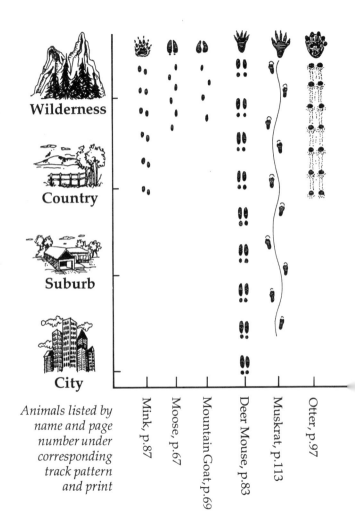

Wilderness

Country

Suburb

City

Animals listed by name and page number under corresponding track pattern and print

Mink, p.87

Moose, p.67

Mountain Goat, p.69

Deer Mouse, p.83

Muskrat, p.113

Otter, p.97

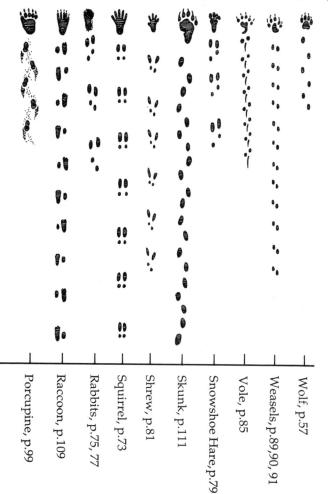

Tracking Notes

Date	Location	Track Pattern	Stride/ Leap
Feb 20, 1989	near path in park	hopper	28 cm

Straddle	Print Measurements	Animal
20 cm	front: L 4 cm 　　　 W 4 cm back: L 12 cm 　　　 W 8 cm	snowshoe hare

Tracking Notes

Date	Location	Track Pattern	Stride/ Leap

Straddle	Print Measurements	Animal

Bibliography

Banfield, A.W.F. 1974. *The Mammals of Canada*. Toronto: University of Toronto Press.

Brown, Tom. 1983. *Tom Brown's Field Guide to Nature Observation and Tracking*. New York: Berkley Books.

Ennion, E.A.R. and N. Tinbergen. 1967. *Tracks*. Oxford at the Clarendon Press.

Forsyth, Adrian. 1985. *Mammals of North America*. Camden House Publishing, Ltd.

Halfpenny, James. 1986. *A Field Guide to Mammal Tracking in Western America*. Boulder: Johnson Books.

Jaeger, Ellsworth. 1948. *Tracks and Trailcraft*. New York: Macmillan.

Lawrence, R.D. 1974. *Wildlife in North America: Mammals*. Thomas Nelson & Sons (Canada) Ltd.

Murie, Olaus, J. 1975. *A Field Guide to Animal Tracks*. 2nd Edition. Boston: Houghton Mifflin Co.

Rue, Leonard Lee. 1981. *Furbearing Animals of North America*. New York: Crown Publishers, Inc.

Rue, Leonard Lee. 1967. *Pictorial Guide to Mammals of North America*. New York: Thomas Y. Crowell Co.

Savage. Arthur and Candace. 1981. *Wild Mammals of Western Canada*. Saskatoon: Western Producer Prairie Books.

Smith, Richard P. 1982. *Animal Tracks and Signs of North America*. Harrisburg, PA: Stackpole Books.

Stokes, Donald and Lillian. 1986. *A Guide to Animal Tracking and Behavior*. Boston: Little, Brown and Company.

Stokes, Donald W. 1976. *A Guide to Nature in Winter*. Boston: Little, Brown and Company.

Webster, David. 1972. *Track Watching*. New York: Frankein Watts, Inc.

Wooding, Frederick. 1982. *Wild Mammals of Canada*. Toronto: McGraw-Hill Ryerson Ltd.

Notes on the bibliography

Distribution map sources: A.W.F. Banfield's *The Mammals of Canada*; F. Wooding's *Wild Mammals of Canada*.

Latin names sources: Hugh Smith, Provincial Museum, Edmonton, AB; F. Wooding's *Wild Mammals of Canada*.

Index

PRESERVING LIFE ON EARTH

The Canadian Parks and Wilderness Society envisages a healthy ecosphere where people experience and respect natural ecosystems. Believing that by ensuring the health of the parts, we ensure the health of the whole, which is our health too.

Join the Canadian Parks and Wilderness Society and you will be helping save endangered spaces right across Canada. You will also receive a subscription to *Borealis* quarterly magazine. Inside you will find dramatic, lively, colorful, hard-hitting coverage of environmental issues in Canada.

To enrol as a member, enclose payment of $31.00 (by cheque, VISA or MC) payable to the Canadian Parks and Wilderness Society. Please provide your full name, address and postal code. If you are paying by VISA or MC, please indicate which card you are using, indicate you are paying $31.00, sign the letter and include your card number and expiry date.

Canadian Parks and Wilderness Society
Suite 1150, 160 Bloor Sreet East
Toronto, Ontario M4W 1B9
(416) 972-0868

CANADIAN PARKS AND WILDERNESS SOCIETY

The Author

Joanne E. Barwise has been an environmental education teacher, outdoor leader, writer and consultant. She co-ordinated the development of a major environmental education centre in Calgary. Her love of the outdoors and teaching children inspired her to create a book about animal tracks. She and her family currently live in Regina.

The Illustrator

Ozlem Boyacioglu has been working as a graphic artist for the past seven years. The majority of her work can be seen in energy education programs. She is studying toward a degree in architecture.